PACKAGING DONE RIGHT IS MAGIC, VOL. I

A Field Guide 33 Years in the Making

Jay Edwards

Pack2Sustain

ISBN 979-8-218-58513-6

Cover art: *Mylar*, oil-on-canvas by skarora.com

For those seeking to move the needle

TABLE OF CONTENTS

Chapter I   **On Shelf: The Promise**
Recycling – Still a Thing
Compostable Packaging – A Virtuous Cycle
Reuse: The Ideal?

Chapter II   **On the Road: The Test**
Data: Hierarchy Kryptonite
Work Hard, Play Hard

Chapter III   **Design & LCA: The Vision**
Purpose Driven
Over? "Easy".
Issue of Material
'Shrooms
Bee Wisdom
Don't Think. Know.

*F*orward

I'm grateful that you have purchased this book! Since you've come this far, it may be as good a place as any to lay out what you can expect as you read on, and maybe I'll do so by sharing with you what this book is not.

This is not a textbook. Textbooks are great for learning the tools to do a job, and your teachers and professors are / were great for challenging you to apply those tools to solve academic problems. The tools and problem-solving skills that you have learned will serve you well, whether your study or work focus is Packaging, Marketing, Food Science, or Design.

Instead, this book is a field guide to the world of Packaging. It will discuss responsibilities you may have, or answers you may need to seek, as you work in the wonderful business of conceiving of a packaging system, putting it to work protecting a product through distribution, and delivering the product / package bundle to the store shelf. Given when this book was written, it won't be a surprise to you that each consideration along the way should be taken through a sustainability lens – and you also won't be shocked at how challenging this can be.

This book will also be the internship you wish you'd had before working full time, the random but impactful conversation with a new colleague that you missed out on at the hotel bar in Chicago or Las Vegas during PackExpo, or the source of a key question or two that helps you navigate a prickly, politically driven issue at work. What you will read are not hot takes – I've had the privilege of working in the Packaging R&D space at both Kraft Foods and as a packaging advisor leading Pack2Sustain, spanning a total of over 30 years. Packaging continues to be an exciting career, and it's time to share what I've learned in the hopes that I might help you prepare for a fulfilling career as well.

You can also expect a strong focus on solutions. Many exciting things are happening in the packaged-product world, and I'll enjoy sharing them with you. (Much work is yet to be done of course, which is convenient since we all like receiving paychecks.)

So, again – thank you! When everything comes together with a packaging project it can indeed be magical, so let's unwind the process starting not from the beginning, but instead from the end.

Chapter I

On Shelf: The Promise

Analogies linking sports and business are pretty common, so let's start with some visualization. As you probably know, in sports visualization is a way for an athlete to rehearse a competition within her mind before taking the field or stepping onto the court. By picturing successful moves or shots, it is believed that athletic performance can be improved when the first whistle blows or the starting gun is fired.

In our packaged product example, I'll challenge you to (figuratively) close your eyes and picture your well-executed product on the shelf of your biggest customer or your favorite retailer. Its presence on the shelf is powerful. Its artwork is inspiring. And everything about it screams "come buy me!" While your goal is obviously to land your product into the shopper's cart, it's also important to remember that you are making a layered and interconnected promise in the process.

At the highest level, you are promising that the shopper – let's call him James – you're promising that James' life will be better in some way with your product in it. Even satisfying a basic need qualifies as making his life better. Getting him out of a jam is an even bigger win and making him feel better about himself is probably the ultimate result. Obviously, each product / package combination is its own unique, complex and nuanced bundle of attributes so I won't try to boil the ocean by attempting to describe the perfect mix. But the one thing almost all product / package bundles have in common is that at some point, the contents of the package are gone, and the package remains. So, let's focus on what we call the package's end-of-life.

To describe what ideal is, it's best to agree on what ideal isn't. We can all probably agree that seeing your brand's package on the side of the road or floating towards you at the beach is not ideal. Most people will feel the same way about seeing the fruit of their labor in a landfill (although you probably don't spend too much time hanging out in landfills). An ideal scenario probably involves your brand's package being repurposed in some way. This allows us to focus on a few scenarios to shoot for.

Recycling – Still a Thing

At the time this book is being written, there is some frustration with recycling, its effectiveness, and its economics. While demand for re-cycled content in packaging continues to increase, getting ahold of quality materials can be challenging. Intermingled material recycling streams are a big contributor to the problem, but there are other complicating factors as well. Not the least of which being the fact that if a body part is found in a load of recyclables, all processing stops, law enforcement is called, and the recycling facility instantly becomes a crime scene.

Shifting – quickly – back to the positive, there is an exciting technology to address many of the less dramatic issues in recycling, and the enabler is one with which you are no doubt getting familiar: artificial intelligence. It may seem strange to associate artificial intelligence with positive development, since concerns about the growth of AI are certainly valid and important. Recycling, however, may prove to be a way for AI to really shine.

The tool I'm speaking of cuts to the heart of the expensive intermingled-recycling-stream issue and does so while sporting a quite dapper name: Greyparrot Analyzer (greyparrot.ai) According to the Greyparrot company's LinkedIn page, "Greyparrot is using AI waste analytics to unlock a new understanding of discarded resources called waste intelligence." The company is essentially building an enormous 'packaging brain' through the various units their customers have installed at recycling facilities around the world. These vision-based, AI-enabled analysis tools can discern a stunning amount of information about nearly every piece of material that flows through a recycling facility – even metrics like mass (without the need to directly weigh the item) and the brand owner (without the need to see the label). As with most AI tools, the more material that flows past the Greyparrot units positioned across the globe, the more accurate these analyses become.

The implications for recycling are significant. The most obvious advantage is superior sorting power, helping operators aggregate cleaner salable products. At the end of the day, the name of the game in recycling is to corral as much marketable material as possible since, again, brand owners are increasingly interested in incorporating recycled content into their packaging. With better analytics and sorting, loads of sorted polypropylene will contain fewer items that are not polypropylene. The same for loads of sorted polyethylene, and for loads of sorted aluminum. All of this adds up to higher profits for recyclers.

What's perhaps less obvious is that recyclers can now pursue a new revenue stream – selling brand-specific recycling data to brand owners. Since very specific collection data is now available, brands that purchase this information can not only show compliance with any relevant Extended Producer Responsibility legislation, but they can also communicate reality-verified recycling rates on their packaging labels. Since Greyparrot Analyzers approximate greenhouse gas impacts, this information can also be purchased by brands for sustainability reporting. Indeed, it's a new day for the humble recycler.

Let's hear from them! The Greyparrot website shares a couple of interesting testimonials, and despite the biased source, they are compelling and worthy of being the last words on the topic for our purposes:

> *"With data gathered throughout the day, it's possible to make informed commercial decisions. We can better understand what we're sending to customers, and decide who our most valuable suppliers are."*
>
> *Mustafa Azimy, Processing Engineer, Suez Recycling & Recovery*

> *"The technology gives us great visibility into our operations and directly aligns with our vision of enabling a circular economy. We're using the systems to provide real-time analysis of the purity of our PET output material."*
>
> *Ian McSpirit, Site Manager, Biffa*

Sticking with quotations, "This is all great," you may be saying, "but what is my role?" As a Packaging Pro, you can help enable recycling of your packaging system through two primary means: design and communication. Good news: there is substantial industry guidance regarding both.

The Association of Plastic Recyclers (plasticsrecycling.org) offers design guidance not only to assist with the selection of plastic materials but also concerning protocols to test and assess these materials for recyclability. Additionally, the APR provides recommendations for testing facilities to perform these assessments, so a reality check of designs you are considering is always within reach.

Plastic packaging can be nuanced. Each plastic resin has its own attributes and drawbacks, and different resins are often used together in a given packaging component to satisfy the requirements of the product it is designed to carry. To call it 'part art, part science' might be a little dramatic, but arriving at the right packaging material for a given application is sometimes complex.

Due to this complexity, plastic packaging tends to dominate the recycling discussion, but similar guidance is also available for other material types. The environmental performance of paper packaging has at least two considerations – proper sourcing and design-for-recycling. The Sustainable Forestry Initiative (forests.org) concerns itself with wood products that are sourced from responsibly managed forests. The SFI provides globally recognized standards to certify these forests, as well as the on-pack labeling to properly communicate the earned certification of the paperboard used to produce cartons and corrugated items.

With responsible sourcing established, paperboard packaging also needs to be designed for recovery. Put differently, it becomes important not to ruin paperboard's inherent recyclability – as the American Forest & Paper Association (afandpa.org) can attest, paper-based packaging is recovered at consistently high rates. There are some design pitfalls to avoid however so becoming familiar and staying current with AFPA guidance is always worthwhile.

Moving on to the metals commonly used in packaging – aluminum and steel – there is guidance here as well. Although recovery rates may vary, the properties of magnetism make these items particularly easy to capture, and the end-markets for packaging metals are well established. The Aluminum Association (aluminum.org) and United States Steel (ussteel.com) provide and track developments for their respective industries. Perhaps the main complexity with recycling metals lies in the use of aluminum foil in flexible packaging applications. Recovery of flexible films is inherently challenging, foil or no foil, but organizations like TerraCycle (terracycle.com) do some admirable work in this space.

Not to be forgotten, glass packaging. The Glass Packaging Institute (gpi.org) is a fine resource for what is perhaps our species' oldest and most storied packaging material.

With any packaging substrate, an understanding of the relevant recyclability facts, objectively determined, is the first step regarding recovery (as mentioned). Communication is the second step, and fortunately there is a tidy resource to help communicate concisely and accurately on pack: How2Recycle (how2recycle.info). By looking at not only the material / packaging-component combination but also across the material recovery chain of custody, How2Recycle can help determine the most appropriate messaging to add to product labels. This is critical to efficiently inform the consumer as to which bin to toss the package into. How2Recycle can even assist with labeling associated with an emerging and controversial category: compostable packaging.

Compostable Packaging – A Virtuous Cycle

"Biodegradable."

"Organic Recycling."

"Anaerobic Digestion."

The vigorous discussion around the concept of used packages just 'going away' makes debates about recycling look tame by comparison. The basic concept is this: a spent package, derived from nature and properly managed after use, is placed into a system where it can return to nature within a reasonable amount of time. Think apple core or banana peel.

It's a very compelling design goal, but not surprisingly several things need to line up for this vision to pan out. Fortunately, the enabling packaging material technologies are out there, and are continually improving. All traditional design demands and parameters – from oxygen barrier to impact resistance – need to be met of course. An additional requirement is that the collection of used packaging must be especially intentional, and its end destination should ideally be uniquely defined. The exciting opportunity is one where a used package can both serve as a nutrient in Nature, and if desired it can also be used to generate more packaging of its own kind.

Things tend to go well with this approach when there is a combination of scale and control. As I have already mentioned and shown, this book has a strong bias towards solutions so I'm pleased to share another one: a late-2023 partnership between the NBA Portland Trail Blazers and foodservice packaging provider Eco-Products (ecoprod ucts.com). There won't be a second strained attempt to apply sports visualization to a business context, because I won't need to. As with the recycling sorting solution example already discussed, these results have entered reality.

The Blazers' home is Moda Center and Veterans Memorial Coliseum, a facility with a stated management goal to only use foodservice packaging that can be recycled, reused or composted. Moda Center is located in Portland, Ore., a metro area that provides one of the United States' most supportive cultures regarding such a goal. Moda is well-suited for the application of compostable packaging for another reason: corralling compostable packaging is much easier when a foodservice operator has a large, 'captive' audience and when it controls the collection systems within its walls.

Once compostable waste is collected at Moda Center after a Blazers game, large concert or other event, it is delivered to Annen Farms, in nearby Mount Angel. The Annen facility is located in Oregon's lovely Willamette Valley, where much of the specialty hops for Portland's world-famous microbreweries are grown, amongst other crops. Annen Farms is also in the composting business, using a technology sourced from Green Mountain Technologies (compostingtechnology.com) to convert 'green waste' generated from their annual harvest into compost that is reapplied to their fields when it's time to plant for the following growing season. The material delivered from Moda Center is added to the feedstock of this compost-producing process.

Satisfying as this real-life virtuous cycle might be, the future could be even better. Women – led and San Francisco – based Mango Materials (mangomaterials.com) has declared that "the future is biodegradable" and their work plays no small part in dragging this dream into existence. This future state starts with methane – usually seen as a problematic waste gas that exacerbates global warming – and ends with the creation of PHA biopolymer. Of course, this is not really the end; the bioplastic resin can then be used to produce packaging components or other items which are themselves biodegradable, ultimately closing the circle back to methane. This is arguably one version of the much-sought Plastic Holy Grail.

That said, let's keep the champagne corked – there is much work yet to be done. While Mango Materials has very deep expertise in PHA, they will be the first to tell you that development of their biopolymer will more than likely be required to satisfy the requirements of a given packaging application. They stand ready to do this work however, and just as petroleum-based plastics currently offer a pallet of options to choose and combine to deliver performance, so too will bioplastics, in time.

As with the process used at Portland's Moda Center, infrastructure is important as well. Unlike folks who might attend a game or concert, consumers of store-bought biodegradable packaging need to be provided with a convenient (as in a really, really convenient) way to put spent packaging 'back' on its way to a facility that can manage the composting process and capture liberated methane. Communities that already offer curbside collection of compostables have a head start on this.

Reuse: The Ideal?

The getting-the-package-back challenge is perhaps even greater for reusable packaging. It's true that long gone are the days of the milk man & and his daily swap of full bottles for empties, but Loop (exploreloop.com) is giving retro a go. Specifically, per their website:

"Loop operates a global reverse supply chain - collecting used packaging from consumers and retailers, enabling deposit return, sorting and storing, and finally returning hygienically cleaned packaging to

manufacturers for refill. We're working hand in hand with industry leaders including Ecolab to guarantee the highest standards of safety at every stage."

This approach allows for a given package – admittedly a beefed-up version relative to its single-use sibling, designed to withstand multiple trips – to theoretically maximize its utility before ultimately giving up the ghost. This may be a good fit for certain products & consumers, especially if retailers can be incentivized to support as a drop-off center to help enable the return journey.

Some brands, like Uni (weareuni.com) hedge their bets a bit and offer a durable 'main' package that can be refilled using flexible film packs. Nearly all the convenience, nearly all the packaging material utilization. This approach can do a nice job of building deep brand loyalty, and again can be a fine fit for the right business.

Chapter II

On The Road: The Test

The last chapter wrapped up with more than a few mentions of challenges and opportunities attached to transporting emptied packages. Now it's time to talk about shipping packages that are full.

Without question, more than delivering clever functionality or displaying handsome graphics, the most important job a package has is to protect its product as it journeys from the filling line to the shopper's cart. If the package fails in this area, nothing else matters. I'm sure that this is not the first place you have heard this, but it bears repeating.

The importance of a packaging system's structural integrity usually leads to a fair amount of testing through the distribution environment, either simulated or via live shipping. This testing is neither glamorous nor easy; it usually requires breaking down until loads (pallets) and / or shipment packaging, hand inspection, notation of leaker frequency and location, and the like. To come away with statistically significant information, large numbers of individual packages may need to be scrutinized, sometimes in less-than-cozy refrigerated – or frozen – distribution centers. Sometimes the leaked product is sticky. Sometimes work 'fires' erupt while you are away doing inspections. But there's only one way to get good data: go get it.

Data: Hierarchy Kryptonite

I know how naive the title of this section sounds, and that I should know better. Organizations of all sizes tend to value some voices more than others. BMoCC (Big Man on Corporate Campus) situations still happen. Organizational charts are still a thing. But packaging is very tangible, and it cares not for titles. One tale in particular will help me to illustrate this.

Before you can be a seasoned Packaging Pro you have to spend time being a newbie, and I was no exception. I began my Packaging R&D career supporting Kraft Food's foodservice business, developing packages for customers that included the restaurant chains Ruby Tuesday's and Subway. The thrill and glamour of seeing one's work on-shelf while grocery shopping was not yet mine to have, but it was interesting to be part of the machinery that made so-called back-of-house operations run.

I did sometimes have the opportunity to 'trade up' to better products, however. When I got wind one fateful day that a Packaging R&D colleague of mine was moving on from her post supporting the foodservice coffee division, I leapt at the chance to fill the void. (A tea drinker in my youth, I had finally and recently discovered coffee and wanted more of it in my life.) Part of this assignment would involve supporting production locations in both the Toronto, ON and Houston, TX areas – this was an additional plus since it would expand my operational exposure to plants outside of the U.S. Midwest.

I was also to join a new cross-functional business team, meet new Marketing folks, and learn new science concerning how the packaging materials and products in my portfolio interact. This team was led by a gentleman with a particular... let's say passion for reducing packaging costs. The phrase that I seem to recall is 'lowest imaginable cost.' And the way that the organization worked at the time, while direct-line reporting relationships existed within each function or department, it was expected that members of this cross-functional team would see its leader as their second boss. (The number of bosses were not limited to two, but we'll keep this story focused.) In this case, we'll call the cross-functional leader Tom.

In all fairness to Tom, he had a boss too, who was also expected to achieve certain metrics, including cost reductions. Eventually Tom's attention came around to the corrugated material used in the shipment packaging for some of our pouched coffee items. A study was requested to investigate material reduction opportunities for these shippers, so I started with lab testing. Based on the test criteria derived from the load and abuse that these shippers were known to experience during warehousing and shipment, the options for corrugated reduction were not performing well. Put simply, the boxes were getting damaged or crushed. These results were communicated to Tom. Tom challenged them.

Ever the diplomat, I reached out to our in-house corrugated material expert – let's call him Chris – and asked him to review my findings and provide his perspective. He did so, and his assessment was the same as mine. Tom rejected both. Chris and I were pressed to perform a live warehouse and shipment test using several until loads of product to fully explore the opportunity.

You hear the term 'fool's errand' every so often, but you don't expect to be the fool. Holding our peace, Chris and I made the phone calls and arrangements we needed to make to get ourselves and some shipper samples to the Houston coffee plant for the decreed test. The plan was to produce and stack several unit loads in the nearby warehouse for 48 hours, prior to shipment and further evaluation. The cases were filled and stacked, and we kept ourselves busy with emails, etc. as the 2 days of stacking began.

We were interrupted about a day and a half later by a phone call from an obviously upset warehouse manager. Mind you, we were in Texas, where folks are not shy about expressing themselves. We were summoned to the scene, which was worse – and more dangerous – than even we imagined. The stacks of coffee that had not already fallen over were at great risk of doing so. We took our photographs and left; it was made clear that we shouldn't plan on coming back soon.

When we returned to Chicagoland HQ, the photographs of the horrifically failed stack test made for a very short conversation with Tom. Time, money and coffee had been wasted, but the larger mission had been accomplished: the futile project was finally canceled. As a bonus, the trip provided a chance for Chris and me to get to know each other over a few good meals and great conversations. Specifically, my cigar diary tells me that we enjoyed Hoyo De Monterrey Excaliburs while solving (the rest of) the world's problems.

Work Hard, Play Hard

Distribution testing isn't always a road to nowhere. Sometimes it can help give the final nod to a packaging project that carries a nice bundle of benefits. The next example also involves foodservice packaging, but the product is not chic coffee but instead mundane mayonnaise.

This project was my first opportunity to lead a cross-functional team towards a technical goal. The goal was simple; achieving it was surprisingly difficult. The team involved a Packaging R&D colleague, both her and my interns, a couple of folks on the Equipment Engineering team, and of course the production plant. The "simple" goal was to pack portion-control pouches of mayonnaise into packs that were narrower – this was an early adoption of the 'stick pack' design approach that has since become very popular.

This configuration has caught on because it helps to optimize the amount of film used for a given volume of product. This saves both material and costs. If done correctly, the number of pouches filled per minute can increase as well, leading to increased operational efficiency. Win, win, win – once you get past the 'Murphies'.

Nothing good ever comes easy, and in this case we were having a hard time keeping mayonnaise out of the pouch seals. Naturally, this problem would cause the pouches to leak. Our Equipment folks beat the bushes trying to understand which filling machine settings were best, and each iteration produced thousands of small mayo pouches just waiting to be gently squeezed upon inspection. Small, smooth explosions were common.

After months of attempts, repeated trips to lovely Champaign, IL for fill tests, and hours upon hours of post-inspection, the team was beginning to get things dialed in. Leaker rates were coming down, and the team felt confident enough to schedule a large run of product to be shipped over the mountains to a distribution center in Stockton, CA. As the trucks traveled on highway I-70 and through the Eisenhower / Johnson Memorial Tunnel, they reached an average elevation of 11,112 feet above sea level, putting the pouch seals to a solid test.

As always, those pouches weren't going to inspect themselves, so the Packaging R&D squad made their travel plans to fly out of Chicago on a nonstop to San Francisco. The vibe was optimism with a touch of anxiety – it wasn't entirely clear how these pouches could have been filled and sealed much better. If this didn't work, the path forward was not clear. The drive east out of San Francisco to Stockton was invigorating; the rental upgrade to a Ford Mustang convertible saw to that. With our target leaker rate defined, it was time to get to work.

Distribution center inspections are tedious, to be sure. Lots of pulling cases off pallets... trying to remember to lift with one's legs... squeezing & inspecting... many corny jokes. All while tabulating the leakers that you find, hoping that they don't exceed a critical agreed-upon maximum number that the business can live with. Our intrepid team endured this tedium, and each other's attempts at humor, and after about 7 hours we consolidated our tallies. Goal: achieved!

The cumulative relief after months of work and pressure was powerful. I phoned the good news back to HQ, and we joyfully hit the highway back to San Francisco, taking a detour through gorgeous Yosemite National Park. A very festive dinner would follow, with a tour of Chandon in Napa Valley the next day. Champagne seemed appropriate.

Chapter III

Design & LCA: The Vision

Let's end this backwards-packaging-project journey at the beginning. A blank slate is at the same time exhilarating and terrifying. There is freedom to create, saddled with the burden of knowing that there are so many ways to get it wrong. Yet those who excel in this space know that the best way to get started is to start.

For packaging, design starts with three main inputs: end-user requirements, material options, and physical form. This order is probably the right priority for these items but following them in the right order is not all the process requires. It's also important to articulate and document the expected environmental impact of each design concept, to further inform the choice. This early stage is where the most opportunity exists to manage and minimize ecological impact;

attempting to 'fix' aspects of a design later on is always a much heavier lift.

This is also the aspect of a career in packaging that can be the most fun. Addressing ambiguity by scouring consumer insights, exploring material suppliers and visiting filling equipment manufacturers certainly delivers on creating a sense of adventure, especially if one such manufacturer is located in Verona, Italy. More on that as I relate an example.

Purpose Driven

A favorite example showing the thrill of intentional design comes, once again, from my experience in Foodservice Packaging R&D. (A retail-market example will follow – I promise.) This example is a favorite not only because of its ultimate success, but because it involved one of the largest and most departmentally diverse project teams that I've ever led. And also because from the very beginning we were focused on a clear physical need.

For those who have never had the pleasure of working behind the scenes in a restaurant, ingredients are received from suppliers in a range of formats and pack sizes, but for items such as condiments they fall mainly into two categories. The first is what is called PC or portion-control pouches (think ketchup or BBQ sauce) and the other is bulk, often in gallon-sized containers. These larger containers are ones that the restaurant patrons may never see, although some restaurants feature them on shelves as a way of enhancing their image through the display of well-known consumer brands (e.g. Hellmann's Mayonnaise.)

While customers may never see most of these bulk containers, restaurant staff surely do, and they must deal with sometimes heavy, awkward, slippery jugs as they dispense product into recipes or smaller containers. Things can get especially messy when the products in question are Thousand Island or French salad dressings. Before moving on to coffee, such products were in my packaging portfolio, and after my marketing colleagues had heard enough complaints from the field, a project was officially launched to design a better bulk bottle for our bulk "pourable" salad dressings.

It is a wonderful thing when a major part of your job involves gazing out a window, pencil in hand, as shapes and design strategies form in your mind. That was a main part of my gig as I drafted proposals to share with the company that supplied the plastic (HDPE) bottles for this project. My initial challenge was to propose designs that not only solved the logistical problems our customers faced but to also get buy-in from the bottle supplier concerning practicality. It made no sense to design a bottle they couldn't make. This dialogue did much to grow my understanding of the HDPE blow molding process and took this business relationship far deeper than the typical transactional level.

Once the overall design direction was set, the usual rigors of preliminary physical testing began, and the business team was invited to weigh in with suggestions regarding enhancements. A new closure needed to be designed to match the new bottle. A completely different labeling approach was required, and more efficient shipment packaging was explored. Much of this work called for close collaboration with some of my favorite colleagues, the Equipment Engineering team.

I loved working with my Equipment team members for a few different reasons. They seemed to carry a cool-older-sibling vibe, in part based on my naive perception that they enjoyed the luxury of operating above the business team political fray. Their focus was on putting the right assets in place to get packages formed, filled, labeled and/or sealed. Period. This work also took them far afield, to locations not limited to product filling plants. For this project, as team lead, I was invited along to a factory acceptance test for the labeling equipment selected for this new bottle, the supplier having their HQ in Verona, Italy. Darn.

Once all was said and done, the team-derived design shown below was awarded a US patent and was so well-received in the field that it won the Association of Dressings & Sauces Package of the Year honor in 2000. There may have been more champagne. As the saying goes, good times.

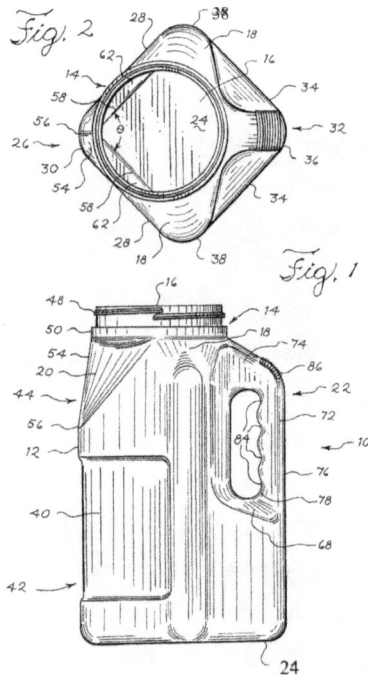

US Patent 6,059,153

Over? "Easy".

I'll take a chance by starting this second example of the joy of 'blue-sky' packaging design work with a statement that might be polarizing: I hate eggs. Their aroma, texture, appearance; can't, stand, them. Unless they are an ingredient in a cookie or piece of cake that I'm about to stuff into my face, I want nothing to do with eggs.

It is ironic then that this particular project had the goals of developing and launching both packaging and product components to deliver a microwavable omelet. The components of an omelet aren't that difficult to assemble; the trick with this concept was how to fold the cooked meal, in the final step of presentation.

Of course, finding a packaging material that could be microwaved was pretty simple – polypropylene (PP) was the obvious place to start. My initial task was to test samples of various shallow PP bowls for microwave performance, using liquid eggs. (This of course exposed me at length to the lovely egg aroma that I so adore.) But the functionality of flipping the egg was elusive.

Sure, the obvious hinged-bottom approach to the PP bowl was the first to be mentioned during project team meetings, but I knew that this would be a bear to execute and might not flip the omelet successfully every time. Then one day, sweet serendipity: as I was cook-testing a bowl, I noticed that the clear sealant layer (facing upwards) had separated / delaminated from the rigid portion below, perhaps due to the heat of the microwave. What if this could be controlled or engineered purposefully, to provide the smooth and seamless flip-functionality we were looking for? What if a small tab could be designed to break off from one side of the bowl, to help facilitate this? One thought led to another, eventually the lawyers got involved (in a good way), and the team ultimately got the novelty-nod from the US Patent Office. Pictures tell the story best:

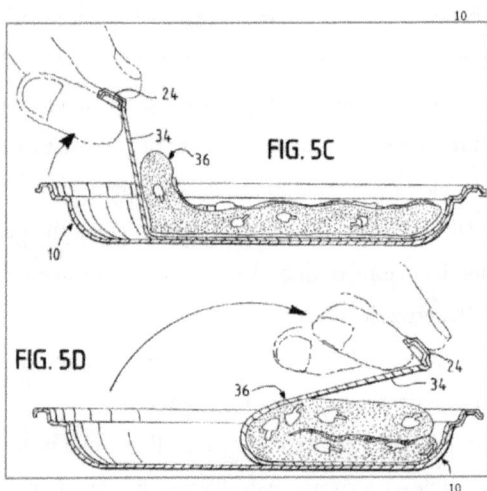

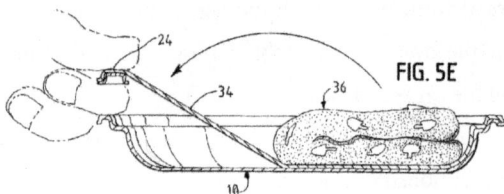

FIG. 5C

FIG. 5D

FIG. 5E

US Patent 0155724 Al

Although I can tell you how this story started, I'm not sure how it ended. I moved on to private advisory work in packaging sustainability shortly after this concept took shape, and since I don't shop for egg products, I'm not sure if it was ever commercialized. Notification through the US Mail that the patent was granted is more than enough.

Issue of Material

My journey into the wide, wonderful world of packaging began with a pivotal choice. While pursuing my Master's Degree in Environmental Engineering from Northwestern, I selected a Material Science elective during my final trimester. That class messed me up. Suddenly, chemistry felt tangibly relevant; it was fascinating to learn why ceramics, metals and plastics behave the way that they do, at the atomic level. When I received an offer from Kraft Foods to work as a Packaging Engineer soon thereafter, I leaped at the opportunity. (A chance to live in Chicagoland during the Michael Jordan era didn't exactly hurt.)

The more I learned about how packaging materials are created and formed, the more my fascination grew. Of course, today's discussions about sustainability often put packaging material choice front and center, with no shortage of debate and controversy. Glass vs. plastic. Post-consumer vs. pre-consumer recycled content. You've seen the posts. It's not always easy to pivot to alternative materials for a whole host of reasons – be they cost, performance, or fear of negative consumer reaction – but I'd like to highlight a couple of interesting approaches that have one thing in common: biomimicry.

'Shrooms

If we are to reinvent packaging materials with the goal of improving their environmental performance, it may be worth looking to the Master for inspiration: Mother Nature. Natural systems have an amazing ability to be resilient without being wasteful; they can be tough, without being toxic. And resources from Nature almost always have the superpower of being renewable.

We need to be careful and specific when we use the term renewable, however. Timing is important. Generally speaking, a natural resource that takes many years to become re-established is less robust than one that only takes a year. One that can regenerate in 7 days is world class. Enter into the chat: mushroom packaging (mushroompackaging.com).

Mushroom packaging supplier Ecovative uses mycelium, the roots of mushrooms, as the biological binder in its process to produce packaging components that are not manufactured, but rather grown. Their main area of focus is protective packaging or 'dunnage' – the packaging within which you might find a new electronic gadget or bottle of wine nestled to keep it safe as it journeys from production or filling plant to its spot on a store shelf. Ecovative's natural approach offers benefits beyond renewability, including flame resistance, high customizability and home compostability. And unlike other materials put to this purpose, it is free of both chemicals and plastics.

We discussed previously how Packaging Job #1 is protecting the product – how great is it to do so using a material that can also reduce environmental impact, when compared to the traditional method? The news gets better; there is a way to achieve the same goal, at the unit load level.

Bee Wisdom

In the discussion of distribution testing back in chapter 2, a few road trips to production plants or distribution centers were mentioned. Usually, the first task after locating test unit load(s) prior to inspection was to borrow a utility knife and cut the plastic stretch wrap away from the neatly organized stacks of shipping containers. This stretch wrap would then be balled up and stuffed into the nearest garbage bin, with focus pivoting to the first of many shippers to be opened and scrutinized. It seemed wasteful, but the unit loads needed to stay secure and organized in shipment, and the stretch wrap achieved this critical goal. Discarding the wrap was taken as an unfortunate means to an end.

It turns out that bees have always known a better way, and humans (the 'inventors' of geometry) are finally catching up. Bees utilize a hexagonal shape to build their hives to store the queen's eggs, pollen and honey. They use this approach because a hexagon requires the least amount of material to hold the most weight. Someone at Hexcelpack (hexcelpack.com) noticed this (likely prior to forming the company.) Utilizing expandable paper to execute this ingenious geometry helps to ensure that their pallet wraps, and other products are recyclable. Further, either virgin or 100% recycled paper can be used as feedstock, enabling another step towards circularity.

Don't Think. Know.

When it comes to sustainable packaging discussions, as with so many other topics, it's easy to get caught up in your feelings. I'm no exception — I have a bias towards technologies that are progressive and innovative, so long as they are also effective. Feelings are horrible tools for making many decisions, however. Decisions require clarity, and clarity requires data.

Authoritative decisions regarding environmental performance of complex packaging options hinge on more than a single metric, and there is a method to appropriately thin the herd. This method is called Life Cycle Analysis, or LCA. There was a strong focus on LCA when sustainable packaging began to come into its own in the U.S. around 2002 or so, but the focus seems to have blurred since. That said, trends tend to cycle and LCA appears to be regaining its rightful place.

Here's why this is so important: across its chain of custody and transformation processes from raw materials, packaging substrates involve and require an array of inputs, resources and by-products, depending on the material in question. As we move from packaging materials to the manufactured packaging components directly used and discarded or repurposed by consumers, this web of resources and energy flows becomes even more complex. The only way to quantify and compare these impacts across packaging materials and design options is through LCA modeling.

As you might guess, reconciling and reporting all this information requires tremendous datasets along with the ability to correctly apply them across the packaging supply chain. I've had the privilege of building or assessing more than a few such LCA tools over the years, and I can easily recommend PIQET (piqet.com) as one worth considering. This tool has a solid grasp of the demand and design nuances pertinent to packaged goods, and outputs a varied list of environmental performance criteria and metrics.

Now, let's manage expectations concerning LCA. While unambiguous numerical reporting provides a great deal of clarity, LCA profiles for packaging options should only be counted on to inform and direct the (many) stakeholder conversations that are needed to arrive at a decision. Ultimately, the core values of a given brand owner will need to be brought to bear, not to mention consumer behavior and the macro dynamics of the relevant marketplace. Best to think of LCA as providing guardrails to keep the team and the business on the right track. Seek progress; don't insist on perfection.

~

Be Well Until Volume II

In the context of consumer goods, packaging is unique. It sits at the crossroads of product development, logistics, on-shelf communication, human consumer experience, and environmental impact. It's no wonder that it can evoke such strong reactions and trigger such inspiration.

I hope you have enjoyed this initial discussion – yes, thoughts are already forming for Volume II of this series. Peace, joy and justice until then!

About the Author

pack2sustain.com/about